International Environmental Labelling

Vol.8 of 11
For All People who wish to take care of Climate Change,
Agriculture & Gardening Industries:
(Shifting Cultivation, Nomadic Herding, Livestock Ranching,
Commercial Plantations, Mixed Farming, Horticulture,
Butterfly Gardens, Container Gardening,
Demonstration Gardens, Organic Gardening)

Jahangir Asadi

Vancouver, BC CANADA

Suggest an ecolabel

If you think that we missed a label and/or you are an ecolabelling body, please consider to submit for the next editions of our 11 Volumes International Eco-labelling Book series. Please send your details, and we'll review your suggestions. Our goal is to be as comprehensive as possible, so thank you for your help!
info@TopTenAward.Net

Copyright © 2022 by Top Ten Award International Network.

All rights reserved. No part of this publication may be reproduced, distributed or transmitted in any form or by any means, including photocopying, recording, or other electronic or mechanical methods, without the prior written permission of the publisher, except in the case of brief quotations embodied in critical reviews and certain other noncommercial uses permitted by copyright law. For permission requests, write to the publisher, addressed "Attention: Permissions Coordinator," at the address below.

Published by: Top Ten Award International Network
Vancouver, BC **CANADA**
Email: Info@TopTenAward.net
www.TopTenAward.net

Ordering Information:
Quantity sales. Special discounts are available on quantity purchases by universities, schools, corporations, associations, and others. For details, contact the "Sales Department" at the above mentioned email address.

International Environmental Labelling Vol.8/J.Asadi—2nd ed.
ISBN 978-1-7773356-2-5

Contents

About TTAIN 10
Introduction 13
General principles of environmental labelling 20
Types of environmental labelling 24
Types I environmental labelling 28
Types II environmental labelling 48
Types III environmental labelling 56
All about 'Eco-friendly' garden 58
Seven ways to create an eco-friendly garden 59
5 supplementary ways to create an eco-friendly garden .. 63
All about 'Eco-friendly' Livestock Ranching 68
How To Create Eco Friendly Butterfly Garden 73
TTAIN Pioneers 80
Bibliography 87
Search by logos 94
Using Algae as Fertilizer 99
Environmental friendly photos 100

I dedicate this book to my son, Taha

We hope that, 10,000 years from now, future generations will be able to see flowers that provide bees with nectar and pollen and... BEES provide flowers with the means to reproduce by spreading pollen from flower to flower...

Jahangir Asadi

Acknowledgements:

I wish to thank my committee members, who were more than generous with their expertise and precious time. I would like to acknowledge and thank the Top Ten Award International Network for allowing me to conduct my research and providing any assistance requested.

It should be noted that all the required permissions for using the logos and trade marks has been obtained to be published in this volume.

About TTAIN

Top Ten Award International Network

Top Ten Award international Network (TTAIN) was established in 2012 to recognize outstanding individuals, groups, companies, organizations representing the best in the public works profession.
TTAIN publishing books related to international Eco-labeling plans to increase public knowledge in purchasing based on the environmental impacts of products.
Top Ten Award International Network provides A to Z book publishing services and distribution to over 39,000 booksellers worldwide, including Apple, Amazon, Barnes & Noble, Indigo, Google Play Books, and many more.
Our services including: editing, design, distribution, marketing
TTAIN Book publishing are in the following categories:

Student
Standard
Business
Professional
Honorary
We focus on quality, environmental & food safety management systems

as well as environmnetal sustain for future kids. TTAIN also provide complete consulting services for QMS, EMS, FSMS, HACCP and Ecolabeling based on international standards.

ISO 14024 establishes the principles and procedures for developing Type I environmental labelling programmes, including the selection of product categories, product environmental criteria and product function characteristics, and for assessing and demonstrating compliance. ISO 14024 also establishes the certification procedures for awarding the label.

TTAIN has enough experiences to help create new ecolabeling programmes in different countries all over the world.
For more detail visit our website : http://toptenaward.net
and/or send your enquiery to the following email:
info@toptenaward.net

CHAPTER 1

Introduction

This book is dedicated to the subject of environmental labels. The basis for the classification of its parts goes back to the types of environmental labelling according to the classifications provided by the International Organization for Standardization. In each section, while presenting the relevant definitions, I mention the existing international standards and present examples related to each type of labelling. Environmental labelling is an important and significant topic, and its richness is added to every day, which has attracted the attention of many experts and researchers around the world. The idea of compiling this book, came to my mind when I observed that national environmental labelling models have been developed in most countries of the world, but in many other countries, the initial steps have not been taken yet. Therefore, I decided to create the first spark for the development of environmental labelling patterns in other countries by collecting appropriate materials and inserting samples of labelling patterns of different countries of the world. It should be noted that the description of each environmental label in this book does not indicate their approval or denial; they are included only to increase the awareness of all enthusiasts and consumers of the meanings and concepts derived from such labels. We hereby ask all interested parties around the world who wish to start an environmental labelling program in their country to benefit from our intellectual assistance and support in the form of consulting contracts. Increasing human awareness of the urgent need to protect the environment has led to changes in all levels of activities, including the production of marketing products, consumption, use, and sale of goods and services at the national and international levels. Stakeholders involved in environmental protection include consumers, producers, traders, scientific and technological institutes, national authorities, local and international organizations, environmental gatherings, and human society in general. Decisions by consumers and sellers of products are made not only on the basis of key points such as quality, price, and availability of

products but also on the environmental consequences of products, including the consequences that a product can have before, after and during production. The most important environmental consequences include water, soil, and air pollution along with waste generation, especially hazardous waste. Further consequences include noise, odor, dust, vibration, and heat dissipation as well as energy consumption using water, land, fuel, wood, and other natural resources. There are further effects on certain parts of the ecosystem and the environment. In addition, the environmental consequences not only include the natural use of the products but also abnormal and even emergency or accidental uses. The basis of studies and studies in this field is done through product life cycle evaluation, which generally involves the study and evaluation of environmental aspects and consequences of a category (product, service, etc.) because of the preparation of raw materials for production until they are used or discarded. Sometimes the phrase "review from cradle to grave" is used for such an evaluation. In addition to the above, the environmental consequences that may occur at any stage of the product life cycle, including the preliminary stages and its preparation, production, distribution, operation, and sale, should also be considered when evaluating it. This type of evaluation refers to product life cycle analysis from an environmental point of view,"

which is a useful tool for measuring the degree of environmental health of a product, comparing different products, improving product quality, and confirming the environmental health claims of the product. The environmental health analysis tool for products and services facilitates their placement in domestic or foreign markets, considering that the awareness of consumers and retailers about the environmental consequences of the product has increased, as has the accurate and explicit measurement by the people in charge at all levels. Local, national, and international in the field of environmental protection. Products that can claim to be environmentally complete in all stages of their life cycle and meet the mandatory and optional environmental needs are considered successful products. Environmental messages refer to the policies, goals, and skills of product manufacturing companies as part of the environmental management systems in which they are applied, and consumers and retailers are increasingly paying attention to this issue when making purchasing decisions. In addition, companies have been encouraged and even forced to adapt their environmental management systems to agencies and retailers and to local, national, international, and other environmental issues.

The environmental health message of a product can be conveyed to the consumer in various ways, including implicitly or explicitly. For example, the implicit or implicit message conveyed directly by the product to the customer is that the product is suitable for the intended use and purpose, and, without material waste in size, weight, and dimensions, is perfectly proportioned and without additional packaging. Sometimes it is necessary to convey these messages and claims about the correctness of the product quite clearly through magazines or other media as well as through certificates that are accurate, simple, and convincing to the consumer in the form of a label. These messages must be accurate and fact-based; otherwise they will nullify the product and create contradictory effects. Confirmation of these claims by a third-party organization will increase its credibility. It should also be noted that the multiplicity of these messages, depending on the type of products or companies producing them, confuses consumers in the market and also creates artificial boundaries or causes a differentiated distinction against certain products or companies. Various models, principles, and methods have been provided by local, regional, national, and international organizations to demonstrate product life cycle analysis and other guidelines on environmental management systems and their labels. At the national level, significant advances have been made in the design of environmental labels in various countries, including developing countries and the Scandinavian countries. For example, the first project was designated in Germany as a Blue Angel in 1977, later on Canada in 1988, the Scandinavian countries and Japan in 1989, the United States and New Zealand in 1990, India, Austria, and Australia in 1991, And in 1992, Singapore, the Republic of Korea, and the Netherlands developed their national environmental labelling. Environmental labels are an environmental management tool that is the subject of a series of ISO 14000 standards. These environmental labels provide information about a product or commodity in terms of its broad environmental characteristics, whether it is about a specific environmental issue or about other characteristics and topics.Interested and pro-environmental buyers can use this information when choosing products or goods. Product makers with these environmental labels hope to influence people's purchasing decisions. If these environmental labels have this effect, the share of the product in question can increase, and other suppliers may create healthy environmental competition by improving the environmental aspects of their products and commodities. The overall goal of environmental labels is to convey acceptable and accurate information that is in no way misleading regarding the environmental aspects of products and commodities, and they encourage the consumer to buy and produce products that reduce stress on the environment. Environmental labelling must follow the general principles that the International Organization for Standardization has published in a collection entitled the ISO 14020 standard, which refers to these general principles here. It should be noted that other documents and laws in this field are considered if they are in accordance with the principles set out in ISO 14020.

What is the purpose of an ecolabel?

What is an ecolabel? Ecolabels are marks placed on product packaging or in e-catalogs that can help consumers and institutional purchasers quickly and easily identify those products that meet specific environmental performance criteria and are therefore deemed "environmentally preferable".

CHAPTER 2

General Principles on Environmental Labelling

1 The First Principle: Evironmental notices and labels must be accurate, verifiable, relevant, and in no way misleading and/or deceptive.

2 The Second Principle: Procedures and requirements for environmental labels will not be ready for selection unless they are implemented by affecting or eliminating unnecessary barriers to international trade.

3 The Third Principle: Environmental notices and labels will be based on scientific analysis that is sufficiently broad and comprehensive, and to support this claim, the product must be reliable and reproducible.

4 The Fourth Principle: The process, methodology, and any criteria required to support the announcements on environmental labels will be available upon request all interested groups.

5 The Fifth Principle: Development and improvement of environmental notices and labels should be considered in all aspects related to the service life of the product.

6 The Sixth Principle: Announcements on environmental labels will not prevent initiative and innovation but will be important in maintaining environmental implementation.

7 The Seventh Principle: Any enforcement request or information requirement related to environmental notices and labels should be limited to the necessary information to establish compliance with an acceptable standard and based on the notification standards and environmental labels.

8 The Eighth Principle: The process of improving the announcement and environmental labels should be done by an open solution with interested groups. Reasonable impressions must be made to reach a consensus through this process.

9 The Ninth Principle: Information on the environmental aspects of the product and goods related to an advertisement and environmental label will be prepared for buyers and interested buyers from a group consisting of an advertisement and an environmental label.

Eco gardening

Eco-gardening is the practise of gardening sustainably, drawing inspiration from nature. It means considering the interactions taking place between the various living elements of the garden: the soil, plants, water... while it encourages the presence of living elements that we consider useful, such as garden allies.

CHAPTER 3

Types of Environmental Labelling

At present, according to the classification provided by the International Organization for Standardization, there are three types of environmental labelling patterns:

1. Type I labelling: This labelling is known as eco-labelling, and because it is difficult to translate this word into many languages, it presents another reason to adhere to a numerical classification system. In the content of Type I labelling, a set of social commitments that creates criteria according to the scientific principles on the basis of which a product is environmentally preferable is discussed. Consumers are then instructed in assessing environmental claims and must decide which packaging is more important.

2. Type II labelling: refers to the claims made on product labels in connection with business centers. This includes familiar claims such as recyclable, ozone-friendly, 60% phosphate-free, and the like. This type of labelling can be in the form of a mark or sentence on the product packaging. Some of them are valid environmental claims—and some can be completely misleading. Usually, all countries have laws against deceptive advertisements, so why has the International Organization for Standardization discussed this issue? The answer is that it is not clear whether the environmental claims have a technical basis or whether the ad is meaningless.

3. Type III labelling: is a distinct form of third-party environmental labelling pattern designed to avoid the difficulties that can result from type-one labelling. Technical committee for Environment of International organization for Standardization has undertaken a new project to standardize guidelines and Type III labelling methods. One of the main objections raised by industries to Type I labelling is the basis for its management.

Seven ways to create an eco-friendly garden:

Recycle and reuse materials. ...
Choose eco materials. ...
Go for local materials. ...
Conserve water. ...
Use permeable paving. ...
Cut back on chemicals. ...
Invest in a green roof. ...

CHAPTER 4

Type I Environmental Labelling

Type I labelling: This labelling is known as eco-labelling, and because it is difficult to translate this word into many languages, it presents another reason to adhere to a numerical classification system. In the content of Type I labelling, a set of social commitments that creates criteria according to the scientific principles on the basis of which a product is environmentally preferable is discussed. Consumers are then instructed in assessing environmental claims and must decide which packaging is more important.

Type I adhesive has the following specifications:
A. Has an optional third-party template.
B. When the product meets a certain standard, the labelling of this product is included.
C. The purpose of this program is to identify and promote products that play a pioneering role in terms of environment, which means its criteria are at a higher level than the average environmental performance.
D. Acceptance/rejection criteria are determined for each group of products and are publicly available.
E. The criteria are adjusted after considering the environmental consequences of the product life cycle.

Examples of Type I Labelling:
In this section, and considering the importance of this type of labelling, I provide a description of some examples of Type I labelling related to some countries along with a list of products on which this mark is placed.

Republic of Korea

The Korea Eco-labelling is a certification system enforced by the Ministry of Environment and KEITI(Korea Environmental Industry & Technology Institute). Since its foundation in April 1992, the system has certified a wide range of eco-friendly products, which were selected as excellent not only in terms of their environmental-friendliness, but also for their quality and performance during their life cycle. Korea Eco-labelling is voluntary certification scheme to attach logo to products with superior environmental quality throughout their lifecycle to other products of the same use, and thus to provide product information to consumers. For 30 years, the scheme has launched plenty of eco-labelling product standards covering personal and household goods, construction materials, office equipment furniture, etc. It products categories which cover all aspects of products, such as reduction of use of harmful substances, energy saving, resource saving, etc. As of April 30th 2021, 169 criterias(=standards), and certifications for 18,250 products(4,549 companies) have maintained.

Contact:
Korea Environmental Industry & Technology Institute(KEITI)
Office of Korea Eco-Label Innovation
Address: 215, Jinheung-ro, Eunpyeong-gu, Seoul, Repulic of Korea
T: +82 2 2284 1518
F: +82 2 2284 1526
E: accolly@keiti.re.kr
W: www.keiti.re.kr

New Zealand

Environmental Choice New Zealand (ECNZ) is the country›s only Government-owned ecolabel. Administered by the New Zealand Ecolabelling Trust, the ecolabel was established in 1992 to provide a credible and independent guide for businesses and consumers to purchase and use products that are better for the environment.

A member of the Global Ecolabelling Network, ECNZ is a Type I ecolabel, which means products and services bearing the label meet criteria covering the whole life-cycle of the product/service, from raw materials, through manufacture and usage, to end-of-life disposal or reuse. Licensed products and services are independently assessed regularly by a third party.

The New Zealand Ecolabelling Trust
PO Box 56 533, Dominion Rd, Mt Eden, Auckland 1446
Tel: 0064 9 845 3330
Email: info@environmentalchoice.org.nz
Web: www.environmentalchoice.org.nz

Peru

BIO LATINA, the consolidated byproduct of four Latin American national certification entities. Since 1998, we have provided certification services in Latin America for national and international markets. We seek to help create a more sustainable and resilient world. With these goals in mind, we have expanded our service portfolio beyond organic to social and environmental certifications.

Visit us: https://biolatina.com

From our regional offices we serve Latin American.

Our headquaters:
Av. Javier Prado Oeste 2501, Bloom Tower Of. 802, Magdalena del Mar, Lima 17, Perú

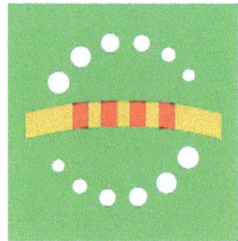

Catalonia

The Emblem of Guarantee of Environmental Quality identifies the products and services that have passed strict environmental quality criteria that go beyond regulatory requirements and bear in mind the life cycle. This type I ecolabelling system, adapted to ISO 14024, is compatible and on a par with other international ecolabelling systems such as the EU Ecolabel.

The Emblem of Guarantee of Environmental Quality was created in November 1994. Its original scope was guaranteeing the environmental quality of certain product properties and characteristics. In 1998, the scope was expanded to include services.

Through the creation of this ecolabel, Catalonia is eager to lead the way in terms of having its own regional ecolabelling system in Europe in keeping with European countries with a long history in environmental protection.

The purpose of the ecolabel is to encourage the design, production, marketing and consumption of more environmentally friendly products and services.

Contact details
Contact person: Josep M. Masip
josepmaria.masip@gencat.cat
ssq.tes@gencat.cat

Bolivia

Legally established in Bolivia, IMOcert has a presence in more than 20 countries in Latin America and the Caribbean, has regional offices in Peru, Paraguay, Mexico, among others. As an organic control body, it has been accredited for many years in accordance with the NOP / USDA Regulation, it also has accreditation of the ISO / IEC 17065 standard and also has other national accreditations of countries where it operates and authorizations for other sustainable schemes and social. IMOcert has extensive experience in certification of producer groups, actively collaborating in the origin and development of the internal control systems methodology.

Complete contact detail
Nombre/Name: Alberto Levy
Gerente Ejecutivo / Executive manager
Teléfono de oficina/office pone: (+591) 4456880/81
Fax: (+591) 44456882
Correo electrónico/ e-mail: imocert@imocert.bio – alevy@imocert.bio
www.imocert.bio

China

China Environmental United Certification Center (CEC), approved by the Ministry of Ecology and Environment of the People's Republic of China (MEE) and accredited by Certification and Accreditation Administration Committee of PRC, is a comprehensive certification and service institution leading in environmental protection, energy saving and low carbon areas. . CEC is committed to serve building national ecological civilization; and has carried out research on environmental protection, energy saving, low carbon development strategies and solutions; has been continuously improving and innovating green industry evaluation system on industrial green development and transition CEC is building a bridge between green production and green consumption by offering independent, impartial and high-quality evaluation and certification service for government, enterprises and the public. CEC is a state-owned, non-profit, legal entity of independent third-party certification. It integrates the certification resource from the former National Accreditation Center for Environmental Conformity Assessment, the Secretariat of China Environmental Labelling Products Certification Committee, Environmental Development Center of MEE, the Chinese Research Academy of Environmental Sciences and other institutions. Business areas includes: products certification, management systems certification, services certification, addressing climate change, energy-saving and energy efficiency certification, green supply chain assessment, environmental stewardship, green credit assessment and green manufacturing system evaluation. CEC also carries out standard establishment and research project and international cooperation and exchanges, etc.

Contact:
Website: http://en.mepcec.com/
E-mail: zhangxiaoh@mepcec.com , zhangxiaoh@mepcec.com

Philippines

The National Ecolabelling Programme Green Choice Philippines (NELP-GCP) is an ecolabelling programme based on ISO 14024 Guiding Principles and Procedures. It is a voluntary, multiple criteria-based, and third-party programme the aims to encourage clean manufacturing practices and consumption of environmentally preferable products and services. It awards the seal of approval to product or service that meets the environmental criteria established for the product category by a multi-sector Technical Committee. Products with the Green Choice Philippines Seal assures the consumers on its preference for the environment. NELP-GCP is being administered by the Philippine Center for Environmental Protection and Sustainable Development, Inc. (PCEPSDI).

Contact:
Website: https://pcepsdi.org.ph/
E-mail: greenchoicephilippines@pcepsdi.org.ph,
greenchoicephilippines@gmail.com

Germany

FSC® is a global not-for-profit organization that sets the standards for responsibly managed forests, both environmentally and socially. When timber leaves an FSC certified forest they ensure companies along the supply chain meet our best practice standards also, so that when a product bears the FSC logo, you can be sure it's been made from responsible sources. In this way, FSC certification helps forests remain thriving environments for generations to come, by helping you make ethical and responsible choices at your local supermarket, bookstore, furniture retailer, and beyond. www.fsc.org

FSC® International
Adenauerallee 134
53113 Bonn
E-mail: info@fsc.org
Phone: +49 (0) 228 367 66

FSC Canada
50 rue Sainte-Catherine Ouest,
bureau 380B, Montreal, QC H2X 3V4
Email: info@ca.fsc.org
Telephone: 514-394-1137

Sri Lanka

National Cleaner Production Centre (NCPC), Sri Lanka was set up by UNIDO in 2002, as a project under the Ministry of Industry to provide the technical expertise and support to the industry and business enterprises in order to prevent pollution and conserve resources by the application of Cleaner Production (CP) and other proactive environmental management tools. NCPC Sri Lanka is registered as a Company by Guarantee not for profit organization under the Act No. 7 of 2007. Over the past two decades, it has evolved as the foremost sustainability solution provider in the country.

The ISO 9001:2015 certified Centre is a registered Energy Service Company (ESCO) under Sustainable Energy Authority (SEA) and a registered consultant under Central Environmental Authority (CEA). It is a founding member of UNIDO/UNEP Resource Efficient and Cleaner Production Network (RECP Net), a global family of 52 NCPCs. NCPC Sri Lanka is a member of Climate Technology Centre & Network (CTCN) and associate member of Global Ecolabelling Network (GEN). Accordingly, we at National Cleaner Production Centre (NCPC), Sri Lanka has developed Eco Labelling scheme under the ISO 14024:2018 - Environmental labels and declarations. NCPC Eco labelling scheme developed, with the Support of United Nations Environment Programme, Under One Planet Network Consumer Information Programme for Sustainable Consumption and Production (CI-SCP).

Contact:
Tel: +94 11 2822272/3,
Fax: +94 11 2822274
E mail: info@ncpcsrilanka.org
Web: www.ncpcsrilanka.org

Hong Kong

The Green Council is a non-profit, tax-exempt charitable environmental stewardship organisation and certification body (Reg. No.: HKCAS-027) of Hong Kong established in 2000. A group of individuals from different sectors of industry and academics shared the vision to help build Hong Kong into a world-class green city for the future. They formed the Green Council with the aim of encouraging the commercial and industrial sectors to include environmental protection in their management and production processes. The Green Council is a non-profit, tax-exempt charitable environmental stewardship organisation and certification body (Reg. No.: HKCAS-027) of Hong Kong established in 2000. A group of individuals from different sectors of industry and academics shared the vision to help build Hong Kong into a world-class green city for the future. They formed the Green Council with the aim of encouraging the commercial and industrial sectors to include environmental protection in their management and production processes. The Green Council is a non-profit, tax-exempt charitable environmental stewardship organisation and certification body (Reg. No.: HKCAS-027) of Hong Kong established in 2000. A group of individuals from different sectors of industry and academics shared the vision to help build Hong Kong into a world-class green city for the future. They formed the Green Council with the aim of encouraging the commercial and industrial sectors to include environmental protection in their management and production processes.

Contact:
Website: https://www.greencouncil.org/hkgls
Email: info@greencouncil.org
Telephone: (852) 2810 1122

ORGANIC CERTIFICATION

Lithuania

EKOAGROS is the only institution in Lithuania for more than 20 years carrying out certification and control activities of organic production and products of national quality, also providing services of certification activities in accordance with the foreign national and private standards in foreign countries. From year 2017 EKOAGROS is accredited as certifying agent to conduct certification activities on crops, wild crops, livestock and handling operations in accordance with USDA NOP.

Contact information:
EKOAGROS
Address K. Donelaicio str. 33, LT-44240 Kaunas, Lithuania
Tel. No. +370 37 20 31 81
Website: www.ekoagros.lt

USA

The Carbonfree® Product Certification is a meaningful, transparent way for you to provide environmentally-responsible, carbon neutral products to your customers. By determining a product's carbon footprint, reducing it where possible and offsetting remaining emissions through our third-party validated carbon reduction projects, companies can:

- Differentiate their brand and product
- Increase sales and market share
- Improve customer loyalty
- Strengthen corporate social responsibility & environmental goals

The Carbonfree® Product Certification Program is proud to be part of Amazon's Climate Pledge Friendly Program!
Carbonfund.org is leading the fight against climate change, making it easy and affordable to reduce & offset climate impact and hasten the transition to a clean energy future.

Contact:

O: 240.247.0630 ext 633
C: 203.257.7808
M: 853 Main Street, East Aurora, NY, 14052

Netherland

For more than 25 years, the independent Dutch foundation SMK works from professional knowledge with companies to improve the sustainability of products and business management. SMK cooperates with an extensive stakeholder network of governments, producers, branch and non-governmental organisations, retailers, consultancies, researchers. The SMK Boards of Experts establish objective criteria for more sustainable products and services. SMK's transparent work processes, third party audits and certifications are conducted according to international certification standards, mostly under supervision of the Dutch Accreditation Council. Besides, SMK is Competent Body of the EU Ecolabel. SMK keeps an extensive database of sustainability criteria.

Contact:
Bezuidenhoutseweg 105 - 2594 AC Den Haag
Telefoon: 070-3586300
Mobiel: 06-82311031
(niet op woensdag)
www.smk.nl

Taiwan

The Green Mark GM) Program was launched by the Environmental Protection Administration of Taiwan (TEPA) in 1992. As the official Type I eco-labeling program, it is in compliance with the requirements of the international stadard, ISO 14024 and is considered an important tool to promote green consumption and production .

To improve the GM application/review mechanism and introduce a third party certification scheme, TEPA promulgated the «Guideline for the Management of Certification Organizations for Environmental Protection Products" in June 2012. Both Environment and Development Foundation (EDF) and the Taiwan Testing and Certification Center (ETC) were commissioned by TEPA as official certifiers. With the expansion of certification capacity and authorization of the certification decision, the certification time was greatly reduced.

Contact :

Website: www.edf.org.tw
TEL: 886-3-5910008 #39
E-mail: lhliu@edf.org.tw

Denmark, Finland

The Nordic Swan Ecolabel
The Nordic Swan Ecolabel is the official Nordic ecolabel supported by all Nordic Governments. It is among the world›s strictest and most recognised environmental certifications.

The Nordic Swan Ecolabel is a Type I environmental labelling program established in 1989 by the Nordic Council of Ministers, connect¬ing policy, people, and businesses with the mission to make it easy to make the environmentally best choice. Nordic Ecolabelling is the non-profit organisation responsible for the Nordic Swan Ecolabel.

The organisation offers independent third-party certification and support for a wide range of product areas and services, ensuring that they comply with the Nordic Swan Ecolabel's strict requirements through documentation and inspections.

30 years of experience and expertise has made the Nordic Swan Ecolabel a powerful tool that paves the way to a sustainable future by giving producers a recipe on how to develop more environmentally sustainable products, and giving consumers credible guidance by helping them identify products that are among the environmentally best.

Globally, you can find more than 25,000 Nordic Swan ecolabelled products. 93% of all Nordic consumers recognise the Nordic Swan Ecolabel as a brand, and 74% believe that the Nordic Swan Ecolabel makes it easier for them to make envi¬ronmentally friendly choices (IPSOS 2019).

Norway, Iceland, Sweden

Securing a sustainable future
The Nordic Swan Ecolabel works to reduce the overall environmental impact from production and consumption and contributes significantly to UN Sustainable Development Goal 12: Responsible consumption and production.
To ensure maximum environmental impact, the Nordic Swan Ecolabel sets product specific requirements and evaluates the environmental impact of a product in all relevant stages of a product lifecycle - from raw materials, production, and use, to waste, re-use and recycling.
Common to all products certified with the Nordic Swan Ecolabel is that they meet strict environmental and health requirements. All requirements must be documented and are verified by Nordic Ecolabelling. Nordic Ecolabelling regularly reviews and tightens the requirements.
Therefore, certifications are time-limited and companies must re-apply to ensure sustainable development.

International website:
Nordic-ecolabel.org
National websites:
Denmark: ecolabel.dk
Sweden: svanen.se
Norway: svanemerket.no   (in Norwegian)
Finland: joutsenmerkki.fi (in Finnish)
Iceland: svanurinn.is   (in Icelandic)

Thailand

The Thai Green Label Scheme was initiated by the Thailand Business Council for Sustainable Development (TBCSD) in October 1993. It was formally launched in August 1994 by The Thailand Environment Institute (TEI) and Thai Industrial Standards Institute (TISI). The Green Label is an environmental certification logo awarded to specific products which have less detrimental impact on the environment in comparison with other products serving the same function. The Thai Green Label Scheme applies to all products and services, but not foods, beverage, and pharmaceuticals. Products or services which meet the Thai Green Label criteria may carry the Thai Green Label. Participation in the scheme is voluntary.

Thailand Environment Institute (TEI)
16/151 Muang Thong Thani, Bond Street,
Bangpood, Pakkred, Nonthaburi 11120 THAILAND
Tel. +66 2 503 3333 ext. 303, 315, 116
Fax. +66 2 504 4826-8
Website: http://www.tei.or.th/greenlabel/
Email: lunchakorn@tei.or.th

EUROPE

Established in 1992 and recognized across Europe and worldwide, the EU Ecolabel is a label of environmental excellence that is awarded to products and services meeting high environmental standards throughout their life-cycle: from raw material extraction, to production, distribution and disposal. The EU Ecolabel promotes the circular economy by encouraging producers to generate less waste and CO_2 during the manufacturing process. The EU Ecolabel criteria also encourages companies to develop products that are durable, easy to repair and recycle. The EU Ecolabel criteria provide exigent guidelines for companies looking to lower their environmental impact and guarantee the efficiency of their environmental actions through third party controls. Furthermore, many companies turn to the EU Ecolabel criteria for guidance on eco-friendly best practices when developing their product lines. The EU Ecolabel helps you identify products and services that have a reduced environmental impact throughout their life cycle, from the extraction of raw material through to production, use and disposal. Recognised throughout Europe, EU Ecolabel is a voluntary label promoting environmental excellence which can be trusted.

Spain , Germany, Italy, Sweden, Greece, Portugal, Poland, Belgium, Netherlands, Estonia, Finland, Austria, Lithuania, Czech Republic, Norway, Cyprus, Ireland, Slovenia, Hungary, Romania, Croatia, Bulgaria, Malta, Slovak Republic, Latvia, Luxembourg, Iceland

Contact and more information via: http://ec.europe.eu

CHAPTER 5

Type II Environmental Labelling

Type II environmental labelling refers to the claims made on product labels in connection with business centers. This includes familiar claims such as recyclable, ozone-free, 60% phosphate-free, and the like. This type of labelling can be in the form of a mark or sentence on the product packaging.
Some of them are valid environmental claims—and some can be completely misleading.

Usually, all countries have laws against deceptive advertisements, so why has the International Organization for Standardization discussed this issue? The answer is that it is not clear whether the environmental claims have a technical basis or whether the ad is meaningless.

Most countries have guidelines at the national level to help producers and consumers know what constitutes a true, scientifically valid claim.
There is a national standard on this in Canada. In Australia, the Consumer Commission has published guidance on this, and there are similar examples in other countries.

Canada

Environmental Sustain for Future kids established in Vancouver, BC Canada in 2020. (ESFK) is an international ecolabel focused on taking care of environment for future of kids.

ESFK defined as 'self-declared' environmental claims made by manufacturers and businesses based on ISO 14020 series of standards, the claimant can declare the environmental objectives and targets in relation to taking care of environment for future kids. However, this declaration will be verifiable.

Environmental Sustain for Future Kids
Vancouver, BC CANADA

Email: info@esfk.org
Web: www.esfk.org

USA

The original recycling symbol was designed in 1970 by Gary Anderson, a senior at the University of Southern California as a submission to the International Design Conference as part of a nationwide contest for high school and college students sponsored by the Container Corporation of America. The recycling symbol is in the public domain, and is not a trademark. The Container Corporation of America originally applied for a trademark on the design, but the application was challenged, and the corporation decided to abandon the claim. As such, anyone may use or modify the recycling symbol, royalty-free.

For More information refer to ISO 14021,
Environmental Labels and declarations

Type III Environmental Labelling

Type III environmental labelling is a distinct form of third-party environmental labelling pattern designed to avoid the difficulties that can result from type I labelling. Technical committee for Environment of International organization for Standardization has undertaken a new project to standardize guidelines and Type III labelling methods. One of the main objections raised by industries to Type I labelling is the basis for its management.

Due to the nature of the system, less than 50% of the various products on the market can meet the criteria and qualify for Type I Labelling. As long as the industry is the main supporter of other third-party models for quality systems, it is sometimes difficult for an industry to support a program that can only benefit 15% of its members. This type of labelling is currently practiced in some countries, such as Sweden, Canada, and the United States. Choosing the right product has never been easy, but Type III labelling will help because each product can have a label that describes its environmental performance and is certified by a third-party company. Consumers can then compare labels and choose their favorite products.

CHAPTER 7

All about 'Eco-friendly' garden

T he environment is – rightly so – on everyone's minds, and if you have a garden, you have an opportunity to contribute to protecting the natural world.

But what does a eco-friendly garden actually look like, and what features does it have? You might imagine an untamed, overgrown jungle teeming with wildflowers and insects.

Nowadays, however, even the slickest contemporary garden designs can be environmentally friendly, thanks to a combination of ethically-sourced materials and innovative technology.

Plan your eco-friendly garden today:

Seven ways to create an eco-friendly garden:

1. Recycle and Reuse Materials
Using recycled materials instead is a great way to go green. The main concern is the origin, extraction, manufacture and installation of materials in structures, paths, walls and patios.

As reclamation yards, especially those in cities, can be expensive, trawl through out-of-town yards and junk shops for materials. New' materials such as paving made from recycled concrete aggregate are now widely available and Many companies now sell pots, fencing and furniture made from recycled wood and plastic.

2. Looking for Ecolabelled Products
Green materials sourced and made by the local community feature strongly in sustainable gardens. Choosing them helps to reduce your carbon footprint as they have few air miles attached, plus most of them use little or no cement, the production of which accounts for more than five per cent of the world's carbon emissions. (Page 36 for more detail). They also give gardens a 'sense of place' by linking them to the local surroundings, which is especially important in rural settings. Materials such as cob (clay and straw), oak, rammed earth, log walls, woven willow, chestnut paling timber and even straw bales are full of character. You will need to consider cost versus durability more keenly than usual, but the suppliers and craftsmen will be able to advise you.

3. Use Local Materials

Choose materials and features, such as paving and pergolas, that have been sourced or built locally, as this will help reduce a garden's carbon footprint and support nearby businesses. For timber products and decking, look for a Forest Stewardship Council (FSC) logo, (Page 32 for more detail) for wood originating from certified plantations.

4. Reduce Water Consumption

Water conservation is essential, so install a butt on every downpipe – you can choose weathered oak barrels or the ubiquitous green plastic tubs. If you have the space, consider an underground rain tank. Larger ones can easily collect enough water for the average garden, plus you can set them up to flush your WC.

Clever cost-cutting irrigation helps too. Don't use a sprinkler on the garden – water the roots of plants without wasting it on the leaves (automated watering systems are useful here); repair leaky pond liners; buy large pots for plants as they don't dry out as quickly; and don't mow the lawn too low in hot weather.

5. Use Premeable Paving

Water run-off from concrete-covered gardens in towns and cities causes localised flooding and affects wildlife significantly. To tackle this problem, legislation has been introduced to regulate the use of solid surfaces in front gardens. You must now use permeable surfacing materials.

Crunchy gravel and slate chippings are the obvious permeable alternative to solid paving, but there are lots of other materials available – from porous asphalt and block paving to grass reinforced with recycled plastic grids.

6. Using Eco Roof

Green roofs are becoming increasingly popular as they help increase biodiversity, provide good insulation, improve air quality and control water runoff – they're also very attractive.

There are plenty of products available using different construction techniques – you can even retrofit an existing shed and garage if they are able to take the weight.

7. Choose Eco-friendly Plants

Choosing the best plants is an important design tool, especially if you're looking to create an organic garden.

In an eco-friendly garden, the best plants will provide food and shelter, creating perfect habitats for beneficial wildlife. Choose lots of local berry-producing plants and trees, such as hawthorn, which might be growing nearby – birds and insects will already be used to them, so they'll visit your garden more frequently if you grow them.

5 supplementary ways to create an Eco-friendly garden:

1. Increase wildlife and biodiversity

Encouraging wildlife makes your garden far more entertaining, as well as helping with pest control – slug-eating hedgehogs and slow-worms love piles of leaves and logs. To attract birds that help with caterpillar control, erect nest boxes and put out a variety of food. Song thrushes love dried fruit, blackbirds adore rotten apples, and sunflower seeds will attract chaffinches and blue tits.

Entice bees by choosing plants with 'open faces'; bright, showy blooms that flower throughout the year. Spring flowers are important for waking bees but are often overlooked; wallflowers, aubrietia and rosemary are all good pollen and nectar sources. For summer, catmint, thyme and lavender are particularly good. Ivy in flower is great in autumn. To help solitary bees, a special bee hotel provides the perfect nesting site.

The key principle to planting is always putting them where they're happiest. Contented plants take care of themselves, but stressed ones need constant feeding and watering, so make sure you don't plant your sun-lovers in the shade, for example, or vice versa. Matching the right plant to the right place will also help keep garden maintenance time to a minimum.

2. Change quality of Soil

Lots of compost and/or well-rotted manure will keep your soil in what gardeners call 'good heart'. This creates a healthy soil teeming with essential micro-organisms, which in turn gives you healthy plants that don't succumb to pests and diseases. Compost soaks up water like a sponge, too – useful in free-draining sandy soils.

Dig in a large bucketful every few feet when planting, or spread liberally around plants as a mulch each spring; this also helps stop light soils from being washed away in heavy rain.

3. Prepare homemade compost

Recycling green waste is important, too – homemade compost costs nothing to make, and will save money on bagged compost and soil conditioner from garden centres.

What to add to your compost:
- veg peelingswood
- hedge trimmings
- grass clippings
- tea bags
- egg boxes
- egg shells
- leaves
- shredded paper
- vacuum cleaner contents

Avoid cooked food, meat, pet faeces and glossy magazines

4. Limit peat-based composts

Peat is dug from peat bogs, causing irreparable damage to precious natural habitats, and as it forms very slowly, simply isn't sustainable either. The problem is that it's great for growing plants as it's sterile, easy to handle, and holds onto nutrients and water like a sponge.

To cut down your use, stop using peat as a mulch or soil conditioner, and instead use **homemade compost**, rotted farmyard manure and leaf mould, which perform better and are full of nutrients. Buy peat-free or reduced-peat compost for potting – it'll say so on the bag – and you'll now find that modern peat-based composts available to gardeners include some of the alternatives, such as bark, wood fibre, coir (coconut husk) or specially formulated green waste.

5. Create a natural garden

Plant according to the garden, not the gardener,' is the ethos at the heart of the natural garden, and looking to nature will provide inspiration and a template to follow. Work with the characteristics of your garden, not against them.

For example, in damp shade, embrace woodland plants and those that grow on woodland margins. For sunny slopes, consider Mediterranean plants like rosemary, juniper, bay and sage – plants with silvery or blue-grey leaves that have naturally adapted to such conditions. Waterlogged soil? Choose wetland plants – not only will your planting visually sit more comfortably, but also promote happy, healthy plants and suffer fewer pests and diseases.

More popular now than ever, native plants are tough, easy to grow and provide food and valuable habitats for wildlife. Ideal for a more relaxed design, they will also help to preserve our threatened plant heritage. Favourites include spiky teasels, gunmetal-coloured cotton thistle, and hardy cranesbills. Quick to colonise poor soils and sunny walls, the red, pink or white flowers of valerian, in particular, are stunning and last for ages. Leave some parts of the garden untidy; nature likes it messy, so gather piles of leaves in undisturbed corners and collect logs and branches, rather than burn them (if they're not diseased). You'll encourage thousands of insects and foraging birds. Hedgehogs also find such spots irresistible to hibernate in.

CHAPTER 8

All about 'Eco-friendly' Livestock Ranching

Ranching is the act of running a ranch, which is essentially an extensive farm for the sole purpose of raising livestock and crops. ... Therefore, we can bring the two definitions together to define livestock ranching as the breeding of animals, for the purpose of food or clothing production. livestock farming, raising of animals for use or for pleasure. ... Ruminant (cud-chewing) animals such as cattle, sheep, and goats convert large quantities of pasture forage, harvested roughage, or by-product feeds, as well as nonprotein nitrogen such as urea, into meat, milk, and wool.

Livestock farmers are facing a number of challenges today. Demands for a lower impact on the environment, especially reducing greenhouse gas emissions, for more animal welfare and for less intensive production need to be balanced with a stable production and a good income. While there are challenges for livestock farmers, there are also many opportunities to increase the resilience and profitability of their farms.

Sustainable pasture management can offer ways to provide good feed to dairy cattle, help reduce livestock emissions, build up and store carbon in the soil, help mitigate the effects of climate change, and much more.

Digitisation and decision support tools can support farmers in better managing their livestock, for instance for more resource- and costefficiency.

Grazing for carbon

Grasslands have enormous potential for storing carbon (C) in the soil. Carbon sequestration improves soil health, makes soils more resilient to extreme weather events, contributes to climate change mitigation and can benefit pasture quality. In sustainable livestock grazing systems, the key challenge is to find the best type of management to combine animal production with soil ecosystem services such as carbon storage, nutrient cycling and biodiversity.

Marketing pasture while reducing emissions

Livestock production significantly contributes to ammonia and greenhouse gas emissions, specifically methane. Adopting grazing methods that let cows graze optimally can contribute directly to lowering livestock emissions.
In addition, there is a growing interest in precision livestock farming, measuring methods, and digital tools that can support farmers in lowering, monitoring or managing farm emissions.

The 'Grazing cow monitor' project has developed a collar that uses GPS tracking to monitor the location of individual cows. The tool clearly indicates how much time a cow spends indoors or outdoors. "The monitor gives farmers digital proof that their cows have spent a sufficient amount of time grazing outdoors", It can therefore help make their administrative tasks easier, but it also allows farmers to label and market their milk as pasture milk." Giving cows access to pastures provides them with fresh grass to eat and can help maintain a healthy soil ecosystem. Outdoor grazing can also help to reduce ammonia emissions from livestock.

When the cows spend less time in the stables, this lowers the chance of faeces reacting with urine and producing ammonia. Keeping track of their cows' pasture time is one of the options for farmers in Flanders and the Netherlands to prove that they are taking measures to lower ammonia emissions.

Farmers can see the results on a dashboard, which they can easily access on their computer or smartphone.

Strategies for managing permanent pasture
Permanent grasslands offer many benefits for biodiversity, ecosystem services such as carbon sequestration, and animal health. Sustainable management strategies can help to maximise these benefits, for instance by matching grassland production with livestock needs.
Digital measuring and decision support tools can help increase resource efficiency and optimise grass production. Differentiating grass-based products such as meat, milk and cheese, can help create higher market value for farmers. Exchanging knowledge is key for increasing profitability, productivity and sustainability for permanent pastures all over the globe..

Robust and resilient dairy farming

Dairy farms are currently faced with economic and environmental challenges, such as volatile prices, extreme weather events, and market demands for more animal-friendly production systems. Improving grazing management can lead to happier cows that produce quality milk with a better price for the farmer. Exchanging experiences can help farmers make their own farms more robust and resilient.

CHAPTER 9

How to create eco friendly Butterfly Garden

Creating a hospitable environment will entice butterflies to stay around long enough to lay eggs for a new generation. By providing the basics of shelter, water, and food—including butterfly-friendly plants—butterflies have a greater chance of thriving and reproducing. Kids love butterflies! Encourage your child's sense of connection to the natural world and invite butterflies into your landscape by planting a butterfly garden. A butterfly garden provides a colorful array of nectar-producing plants that not only attract butterflies (and often hummingbirds as well), but offers plants to feed the caterpillar stage of their life cycle. With the appropriate plantings, a butterfly garden provides opportunities to educate your children about the life cycle of a butterfly, allowing them to view each stage of growth and explore the intricate relationships of plants and animals.

Materials:
- in-ground garden space, raised bed or container garden
- trowel or shovel
- flowering plants for adult butterfly
- host plans for caterpillars

Approximate Time to Complete: 3 to 4 hours to plan, gather plants and install; multiple weeks to grow and attract butterflies

Location: Outdoor
Ages: fun for all ages
Season: spring through fall

Start with a blank slate by pulling out dead bush, removing the plants that were overgrown, and just getting down to the bare earth.

Select plants that grow well in your area. You need to include flowering plants that attract butterflies (many butterflies have favorite plants to sip from) and also leafy "host plants" that attract egg-laying butterflies and provide food for the caterpillars (also known as the larvae). It is always best to select native plants that will attract butterflies native to your area.

Preferred Plants For Caterpillars
Note: Planting trees and taller bushes will help give your garden the sustainability it needs to feed your caterpillars and preserve the flowers for your butterflies.

Trees
Serviceberry, River Birch, Hackberry, Kousa Dogwood, Apple, Cherry, Plum, Willow

Here are a few examples of common butterflies and their preferred food sources:

Butterfly	Host plant(s) for caterpillars	Nectar plants for adult butterflies
Monarch	Milkweed	Milkweed, asters, red clover, zinnia, cosmos, lantana, pentas, daisy
Eastern Black Swallowtail	Carrots, celery, dill, parsley, Queen Anne's lace, rue, Texas turpentine broom	Milkweed, phlox
Giant Swallowtail	Citrus, hop tree, prickly ash, rue	Lantana, orange tree
American Painted Lady	Daisies, everlastings, and other composites	Burdock, daisy, everlastings, mallow, yarrow, zinnia, heliotrope
Orange Sulphur	White clover, alfalfa, vetch, lupine	Clovers, dandelion, parsley, zinnia, other meadow flowers, members of the composite family
Silver-Spotted Skipper	Beans, beggar's tick, licorice, locusts, wisteria	Many garden and meadow flower
Variegated Fritillary	Violets, pansies, stonecrops, passionflowers	Meadow flowers, hibiscus, composite family

Plant your garden. Add one or two large, flat rocks in the sun so butterflies a place to bask when mornings are cool. Since butterflies cannot drink from open water, provide them with a "puddle" by filling a container, such as an old birdbath, with wet sand where they can perch and drink safely.

Once the garden is planted, stand back and wait for the butterflies to stop by. With a successful butterfly garden, your kids will be able to observe the developmental process of a butterfly.

The eggs soon hatch, and the larvae appear and eat the leafy growth of the host plant, eventually developing into full-grown caterpillars. Remember, you will need to tolerate some leaf damage from your very hungry garden guests.

Later, these caterpillars affix themselves to a twig or branch and form a chrysalis, entering the pupa stage. Within about two weeks, they metamorphose into butterflies and re-emerge. Avoid all pesticides. Butterflies are insects, so it makes sense that insecticides — even those labeled organic" — can harm them.

Consumers are increasingly demanding sustainable products that are not toxic to themselves or the environment. The natural market is growing exponentially, and choosing raw, natural materials will cement your brand as a safe choice — both environmentally and economically

CHAPTER 10

Top Ten Award International Network Environmental Pioneers

Top Ten Award international Network (TTAIN) was established in 2012 to recognize outstanding individuals, groups, companies, organizations representing the best in the public works profession. TTAIN publishing books related to international Eco-labeling plans to increase public knowledge in purchasing based on the environmental impacts of products. We introduce in each volume some of the organizations that are doing their best in relation to taking care of the environmnet.

Germany

FSC® is a global not-for-profit organization that sets the standards for responsibly managed forests, both environmentally and socially. When timber leaves an FSC certified forest they ensure companies along the supply chain meet our best practice standards also, so that when a product bears the FSC logo, you can be sure it's been made from responsible sources. In this way, FSC certification helps forests remain thriving environments for generations to come, by helping you make ethical and responsible choices at your local supermarket, bookstore, furniture retailer, and beyond. www.fsc.org

FSC® International
Adenauerallee 134
53113 Bonn
E-mail: info@fsc.org
Phone: +49 (0) 228 367 66

FSC Canada
50 rue Sainte-Catherine Ouest,
bureau 380B, Montreal, QC H2X 3V4
Email: info@ca.fsc.org
Telephone: 514-394-1137

Note:
We've done our absolute best to provide the best information possible, but since we haven't tried every single one of these solutions in every possible cleaning situation, we can't vouch for them 100 percent.

UNEP

The United Nations Environment Programme (UNEP) is the leading global environmental authority that sets the global environmental agenda, promotes the coherent implementation of the environmental dimension of sustainable development within the United Nations system, and serves as an authoritative advocate for the global environment.

Our mission is to provide leadership and encourage partnership in caring for the environment by inspiring, informing, and enabling nations and peoples to improve their quality of life without compromising that of future generations.

Headquartered in Nairobi, Kenya, we work through our divisions as well as our regional, liaison and out-posted offices and a growing network of collaborating centres of excellence. We also host several environmental conventions, secretariats and inter-agency coordinating bodies. UN Environment is led by our Executive Director.

We categorize our work into seven broad thematic areas: climate change, disasters and conflicts, ecosystem management, environmental governance, chemicals and waste, resource efficiency, and environment under review. In all of our work, we maintain our overarching commitment to sustainability.

Website: www.unep.org

Bibliography

Bibliography:

Amberg, N.; Magda, R. Environmental Pollution and Sustainability or the Impact of the Environmentally Conscious Measures of International Cosmetic Companies on Purchasing Organic Cosmetics. Visegrad J. Bioecon. Sustain. Dev. 2018, 1, 23.

Asadi, J., "International Environmental Labelling, Economic Consequencies, Export Magazine, July 2001

Asadi, J. 2008. Mobile Phone as management systems tools, ISO Magazine, Vol.8, No.1

Asadi, J., Eco-Labelling Standards, National Standard Magazine, Sep. 2004.

Barbieux, D.; Padula, A.D. Paths and Challenges of New Technologies: The Case of Nanotechnology-Based Cosmetics Development in Brazil. Adm. Sci. 2018, 8, 16.

CHOI, J.P. Brand Extension as Informational Leverage. Review of Eco- nomic Studies, Vol. 65 (1998), pp. 655-669.

Corrado, M., (1989), The Greening Consumer in Britain, MORI, London

Corrado, M., (1997), Green Behaviour – Sustainable Trends, Sustainable Lives?, MORI, london, accessed via countries. Manila, Asian Development Bank 33p.

Cosmetics, Perfume, & Hygiene in Ancient Egypt. Available online: https://www.ancient.eu/article/1061/cosmetics-perfume--hygiene-in-ancient-egypt/

He Z, Xu J X 1993 Evalustion and measurement of Landscape greening benefit J.Chinese Landscape Architecture. 03 46-51

Davies, Clive. Chief, Design for the Environment Program, EPA March 24, 2009.

EIP-AGRI network at www.eip-agri.eu

The Family Butterfly Book by Rick Mikula. Storey Publishing,

Federal Trade Commission, "Sorting Out Green Advertising Claims." http://www.ftc.gov/bcp/edu/pubs/consumer/general/gen02.shtm (March 26, 2009, March 27, 2009)

MSNBC, "Do You Know What's in Your Cleaning Products?" http://today.msnbc.msn.com/id/29663739/ (March 17, 2009)

Ooyen, Carla. Research Manager with Nutrition Business Journal. Personal correspondence. March 19, 2009.

Tekin, Jenn. Marketing Manager with Packaged Facts & SBI. Personal correspondence. March 17, 2009.

University of California - Berkeley. http://berkeley.edu/news/media/releases/2006/05/22_householdchemicals.shtml (March 26, 2009)

Feenstra, R.C. "Exact Hedonic Price Indexes," Review of Economics and Statistics 77 (1995): 634-653.

Feenstra, R.C., and J.A. Levinsohn. "Estimating Markups and Market Conduct with Multidimensional Product Attributes," Review of Economic Studies (62 (1995): 19-52.

Forest Stewardship Council: "Principles and criteria for forest stewardship" Document 1.2: <http://www.fscoax.org>

Forsyth, K. 1999. Will consumers pay more for certified wood products? Journal of Forestry 97 (2) : 18-22.

Freeman, A. M III. The Measurement of Environmental and Resource Values. Theory and Methods. Washington D.C.: Resource for the Future, 1993.

Friends of the Earth, 1993. Timber certification and eco-labeling. London, FOE:

Geetha Margret Soundri, "Ecofriendly Antimicrobial Finishing of Textiles Using Natural Extract", Journal of International Academic Research For Multidisciplinary, ISSN: 2320 – 5083, 2014, Vol 2.

Graves, P., J.C. Murdoch, M.A. Thayer, and D. Waldman. "The Robustness of Hedonic Price Estimation: Urban Air Quality," Land Economics 64(1988): 220-233.

Halvorsen, R. and R. Palmquist. "The Interpretation of Dummy Variables in Semilogarithmic Equations." American Economic Review 70:474-75 (1980).

Imhoff, Dan, and Grose, Lynda, and Carra, Roberto., "Organic Cotton Exhibit," Mimeo. Simple Life and distributed the Texas Organic Cotton Marketing Cooperative, O'Donnell, Texas (1996).

Imhoff, Dan. "Growing Pains: Organic Cotton Tests the Fibre of Growers and Manufacturers Alike," reprinted on Simple Life's web page (simplelife.com), but first printed by Farmer to Farmer, December 1995.

IISO 14020, ISO 14021,ISO 14024,ISO 14025, International Organization for Standardization.

Kennedy, P.E. "Estimation with Correctly Interpreted Dummy Variables in Semilogarithmic Equations," American Economic Review 71: 801 (1981).

Kirchho®, S., (2000), Green Business and Blue Angels.

Labeling Issues, Policies and Practices Worldwide.

Lamport, L. 1998. The cast of (timber) certifiers: who are they? International J. Ecoforestry 11(4): 118-122.

Large Scale impoverishment of Amazonian forests by logging and fire. 1999.

Lathrop, K.W. and Centner, T.J. 1998. Eco-labeling and ISO 14000: An analysis of US regulatory systems and issues concerning adoption of type II standards. Environmental

Lee, J. et al. 1996. Trade related environmental measures; sizing and comparing impacts.

Lehtonen, Markku. 1997. Criteria in Environmental Labeling: A comparative Analysis on Environmental Criteria in Selected Labeling Schemes. Geneva, UNEP. 148p.
LIEBI, T. Trusting Labels: A Matter of Numbers? Working Paper Uni versity of Bern, No. 0201 (2002).
Lindstrom, T. 1999. Forest Certification: The View from Europe's NIPFs. Journal of Forestry 97(3): 25-31. London
Losey, J.E., Rayor, L.S. & Carter, M.E. 1999. Transgenic pollen harms monarch larvae. Nature 399 20 May): p.214.
Management 22 (2) : 163-172.
Mattoo, A. and H. V. Singh, (1994), Eco-Labelling: Policy Considera-Michaels, R. G., and V. K. Smith. "Market Segmentation And Valuing Amenities With Hedonic Models: The Case Of Hazardous Waste Sites," Journal of Urban Economics, 1990 28(2), 223-242.
Nicholson-Lord, D., (1993) 'Tis the Season to be Green, The Independent, 20 December
Nuttall, N., (1993), Shoppers can cross green products off their lists, The Times, 3 July
OCDE/GD(97)105. Paris, OECD. 81p.
OECD. "Ec-labelling: Actual Effects of Selected Programmes," OCDE/GD (97) 105, 1997, Paris. (available on line at http://www.oecd.org/env/eco/books.htm#trademono)
OECD. 1997a. Case study on eco-labeling schemes. Paris, OECD (30 Dec):
OECD. 1997b. Eco-labeling: Actual Effects of Selected Programs.
Osborne, L. "Market Structure, Hedonic Models, and the Valuation of Environmental Amenities." Unpublished Ph.D. dissertation. North Carolina State University, 1995.
Osborne, L., and V. K. Smith. "Environmental Amenities, Product Differentiation, and market Power," Mimeo, 1997.
Ozanne, L.K. and Vlosky, R.P. 1996. Wood products environmental certification: the United States perspective". Forestry Chronicle 72 (2) : 157-165.
Palmquist, R. B., F. M. Roka, and T.Vukina. "Hog Operations, Environmental Effects, and Residential Property Values," Land Economics 73(1), (1997): 114-24.
Palmquist, R.B. "Hedonic Methods," in J.B Braden and C.D. Kolstad, eds. Measuring the Demand for Environmental Improvement. Amsterdam, NL: Elsevier, 1991.
Pento, T. 1997. Implementation of Public Green Procurement Programs (22-31) in Greener Purchasing: Opportunities and Innovations. Sheffield, Greenleaf Publ. 325 p.
Polak, J. and Bergholm, K. 1997. Eco-labeling and trade: a cooperative approach (Jan.): Policy in a Green Market. Environmental and Resource Economics 22, 419-
Poore, M.E.D. et al. 1989. No timber without trees. London, Earthscan. 352p.

Raff, D. M.G., and M. Trajtenberg. "Quality-Adjusted Prices for the American Automobile Industry: 1906-1940." NBER Working Paper Series, Working Paper No. 5035, February 1995.

Roberts, J. T. 1998. Emerging global environment standards: prospects and perils. Journal of Developing Societies 14 (1): 144-163.

Rosen, S., "Hedonic Prices and Implicit Markets: Product Differentiation in Pure Competition." Journal of Political Economy. 82: 34-55 (1974).

Ross, B. 1997. Eco-friendly procurement training course for UN HCR. : 126 p.

Salzman, J. 1997. Informing the Green Consumer: The Debate over the Use and Abuse of Environmental Labels. Journal of Industrial Ecology 1 (2): 11-22.

Sanders, W. 1997. Environmentally Preferable Purchasing: The US Experience (946-960) in Greener Purchasing: Opportunities and Innovations. Sheffield, Greenleaf Publ. 325p.

Sayre, D. 1996. Inside ISO 14000: The competitive advantage of environmental management. Delray Beach FL., St. Lucie Press. 232p.

SHAPIRO, C. Premiums for High Quality Products as Returns to Reputa- tion. Quarterly Journal of Economics, Vol. 98, No. 4 (1983), pp. 659-680.

Stillwell, M. and van Dyke, B. 1999. An activists handbook on genetically modified organisms and the WTO. Washington DC., The Consumer's Choice Council: 20 p.

Semenzato, A.; Costantini, A.; Meloni, M.; Maramaldi, G.; Meneghin, M.; Baratto, G. Formulating O/W Emulsions with Plant-Based Actives: A Stability Challenge for an Eective Product. Cosmetics 2018, 5, 59.

Teisl, M. F., B. Roe, and R. L. Hicks. "Can Eco-labels tune a market? Evidence from dolphin-safe labeling," Presented paper at the 1997 American Agricultural Economics Association Meetings, Toronto.

www.futureplc.com

He F C, Li Y, Yang X M, Qin F 2017 Exploration on the development path of old industrial base transition:taking Xuzhou city as an example to establish a national Eco-garden city J. Chinese Landscape Architecture. 33 91-95

Tibor, T. and Feldman, I. 1995. ISO 14000: a guide to the new environmental management standards. Burr Ridge Ill., Irwin Professional Publ. 250 p.

Du L P 2009 Some thoughts on the present situation of the application of ground cover plants J. Northern Horticulture. 08 229-231

Du H Z, Song W 1996 A brief analysis of ecological benefits of greening in Song Qingling cemetery J. Journal of Jiangsu Forestry & Technology. S1 88-91

Townsend, M. 1998. Making things greener: motivations and influences in the greening of manufacturing. Aldershot, England, Ashgate Publisher. 203p.

U.S. Energy Information Administration, What is U.S. Electricity Generation by Energy Source?, Retrieved From: https://www.eia.gov/tools/faqs/faq.php?id=427&t=3

U.S. Energy Information Administration, Biomass Explained, Retrieved From: https://www.eia.gov/energyexplained/?page=biomass_home

U.S. Environmental Protection Agency. National Water Quality Fact Inventory: 1990 Report to Congress. EPA 503-9-92-006, Apr. 1992.

UK Eco-labelling Board website, accessed via http://www.ecosite.co.uk/Ecolabel-UK/

US EPA, 1993. Determinants of effectiveness for environmental certification and labeling programs. Washington, D.C., US Environmental Protect

US EPA, 1993. Status report on the use of environmental labels worldwide. Washington, D.C., US Environmental Protection Agency (742-R-93-001 September).

US EPA, 1993. The use of life-cycle assessment in environmental labeling. Washington, D.C., US Environmental Protection Agency (742-R-93-003 September).

Wang L Y 1994 How to improve the ecological benefits in the construction of Jinan garden J. Chinese Landscape Architecture. 03 56-58+55

US EPA, 1998. Environmental labeling: issues, policies, and practices worldwide.

US EPA, 1999. Comprehensive procurement guidelines (CPG) program. Washington, D.C., US Environmental Protection Agency: <www.epa.gov/cpg>

US EPA, 1999. Environmentally preferable purchasing program: Private sector pioneers: How companies are incorporating environmentally preferable purchases. Washington, D.C.,

USG, 1993. Federal acquisition, recycling, and waste prevention. Washington DC., Executive Order: (20 October).

USG, 1998. Greening the government through waste prevention, recycling, and federal acquisition. Washington, D.C., Executive Order 13101 (September).

Kijjoa, A.; Sawangwong, P. Drugs and Cosmetics from the Sea. Mar. Drugs 2004, 2, 73–82. [CrossRef]

Wang, J.; Pan, L.; Wu, S.; Lu, L.; Xu, Y.; Zhu, Y.; Guo, M.; Zhuang, S. Recent Advances on Endocrine Disrupting Eects of UV Filters. Int. J. Environ. Res. Public Health 2016, 13, 782.

Bilal, A.I.; Tilahun, Z.; Shimels, T.; Gelan, Y.B.; Osman, E.D. Cosmetics Utilization Practice in Jigjiga Town, Eastern Ethiopia: A Community Based Cross-Sectional Study. Cosmetics 2016, 3, 40.

Ting, C.T.; Hsieh, C.M.; Chang, H.-P.; Chen, H.-S. Environmental Consciousness and Green Customer Behavior: The Moderating Roles of Incentive Mechanisms. Sustainability 2019, 11, 819.

Chen, K.; Deng, T. Research on the Green Purchase Intentions from the Perspective of Product Knowledge. Sustainability 2016, 8, 943.

Wang, H.; Ma, B.; Bai, R. How Does Green Product Knowledge Eectively Promote Green Purchase Intention? Sustainability 2019, 11, 1193.

Nguyen, T.T.H.; Yang, Z.; Nguyen, N.; Johnson, L.W.; Cao, T.K. Greenwash and Green Purchase Intention: The Mediating Role of Green Skepticism. Sustainability 2019, 11, 2653.

Cinelli, P.; Coltelli, M.B.; Signori, F.; Morganti, P.; Lazzeri, A. Cosmetic Packaging to Save the Environment: Future Perspectives. Cosmetics 2019, 6, 26.

Eixarch, H.; Wyness, L.; Siband, M. The Regulation of Personalized Cosmetics in the EU. Cosmetics 2019, 6, 29.

Appendix I: Search by Logos

Here you can search the logos in this volume. It will help you to better undersand the Ecolabels you may encounter while shopping. Buying Eco-products will aid in having a better environment with minimum polution during production processes. Three important parameteres for shopping are **quality**, **price** & **environmental impacts** of the products.

INTERNATIONAL ENVIRONMENTAL LABELLING VOL.8 • 95

Vol.8　Goto page:　41	Vol.8　Goto page:　46
Vol.8　Goto page:　50	Vol.8　Goto page:　38
Vol.8　Goto page:　39	Vol.8　Goto page:　30
Vol.8　Goto page:　37	Vol.8　Goto page:　42

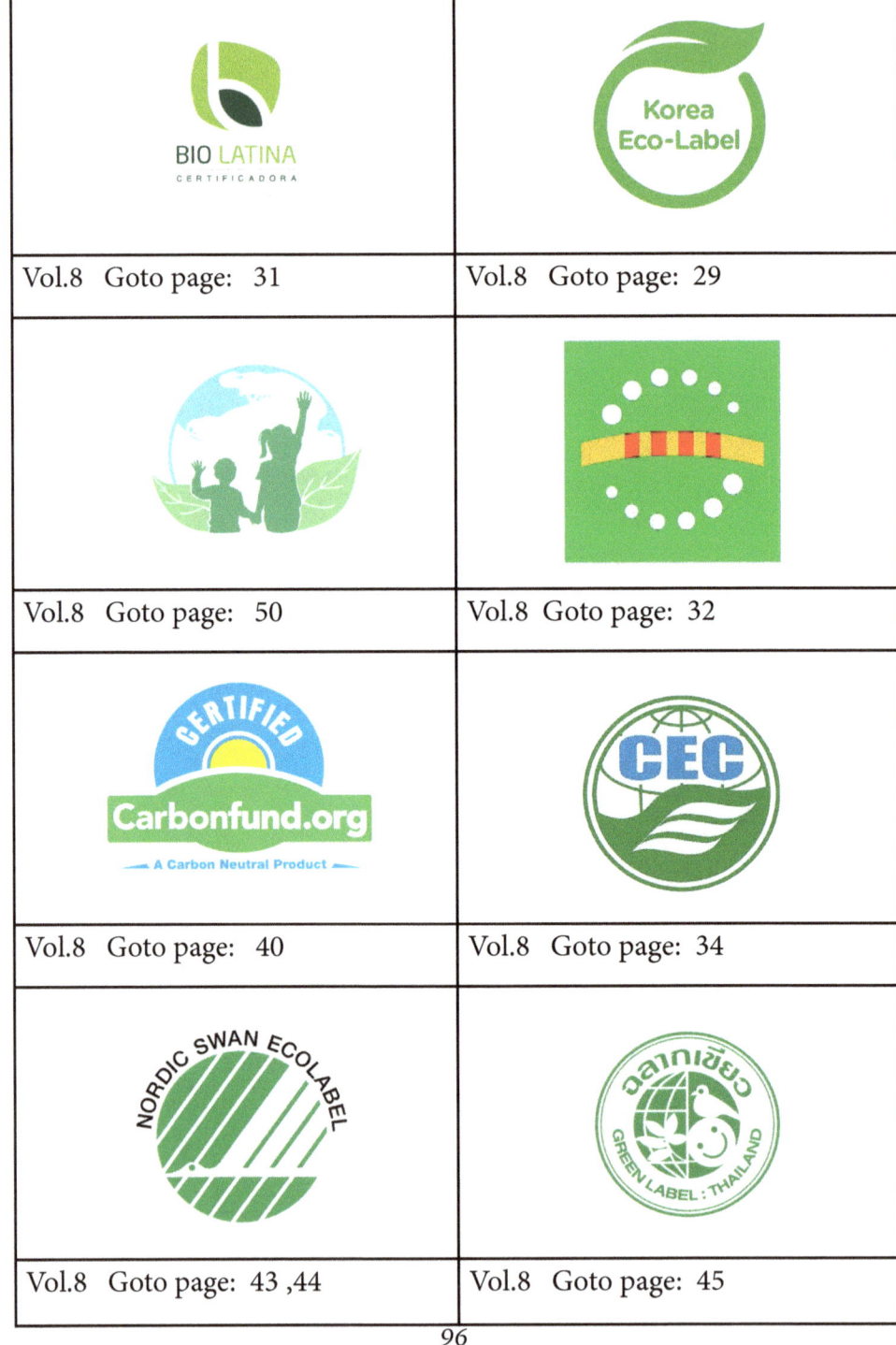

INTERNATIONAL ENVIRONMENTAL LABELLING VOL.8

Vol.8　Goto page: 41	Vol.8　Goto page: 46
Vol.8　Goto page: 33	Vol.8　Goto page: 36
Vol.8　Goto page: 39	Vol.8　Goto page: 30
Vol.8　Goto page: 37	Vol.8　Goto page: 42

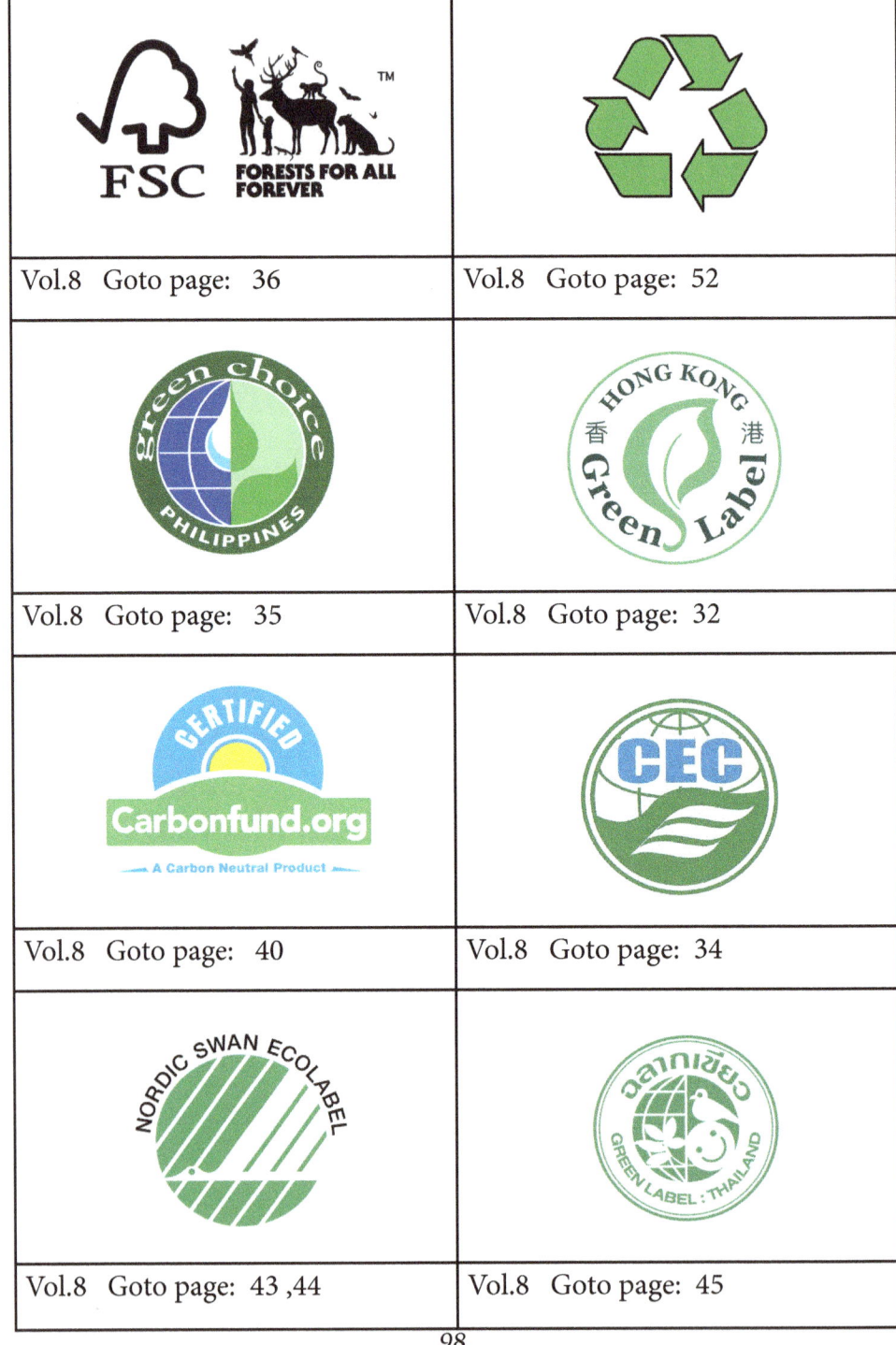

Vol.8　Goto page:　36	Vol.8　Goto page:　52
Vol.8　Goto page:　35	Vol.8　Goto page:　32
Vol.8　Goto page:　40	Vol.8　Goto page:　34
Vol.8　Goto page:　43 ,44	Vol.8　Goto page:　45

Appendix II

Using Algae as Fertilizer

Algae fertilizer is especially useful because it is a living organism. When you use algae as a fertilizer, it quickly begins to break down releasing its abundant nitrogen source. As a result, algae can act as an excellent fertilizer that can be used to help grow crops efficiently.

The overall goal of the project AlgalFertilizer was to assess the feasibility of using green algae to extract phosphorus from waste-water streams and to apply the phosphate-rich algal biomass directly as a fertilizer to the soil. The motivation for this research is the fact that phosphate is currently extracted from finite rock deposits and is very inefficiently used. Therefore, the project contributes to find sustainable strategies to manage the global phosphate reserves more efficiently.In four Work Packages we investigated different aspects of including algae in the phosphate usage cycle. We investigated the efficiency of phosphate uptake by selected green algae, studied how the phosphate is taken up and converted by the algae, monitored how the phosphate is released from the algae to the soil when applied as fertilizer and how the phosphate enters the plant, and we built mathematical models describing these processes to gain a theoretical understanding of the underlying mechanisms and to provide the basis to built predictive tools that will

help to optimize the algal fertilizing strategy. We found that algae are capable of taking up by far more phosphate than they need to grow. Specifying the exact conditions and identifying the best strains for such a luxury uptake will now help to optimize phosphate extraction of waste-water streams, thus making phosphate use more efficient and contributing to waste-water treatment simultaneously. Internally algae store phosphate in the form of inorganic poly-phosphate. To quantify and localize these phosphate pools we have developed a novel technique based on Raman spectroscopy.

Appendix III

Environmental Friendly Photos

Environmental friendly photos will be placed in this appendix. These photos can be received in the Top Ten Award International Network inbox from anywhere and everywhere, all over the globe. You can send your appropriate photos to us for them to be considered for publishing in one of the future, related volumes. They will be published with proper credit to the sender. The pictures can also be images of the Ecolabels existing in products within your country.

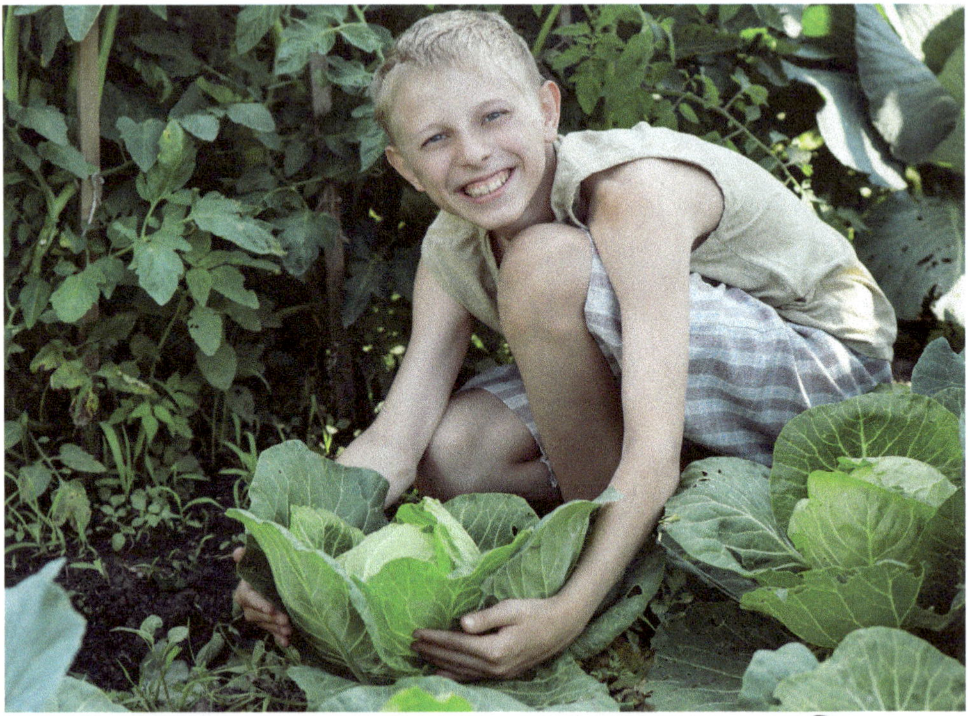

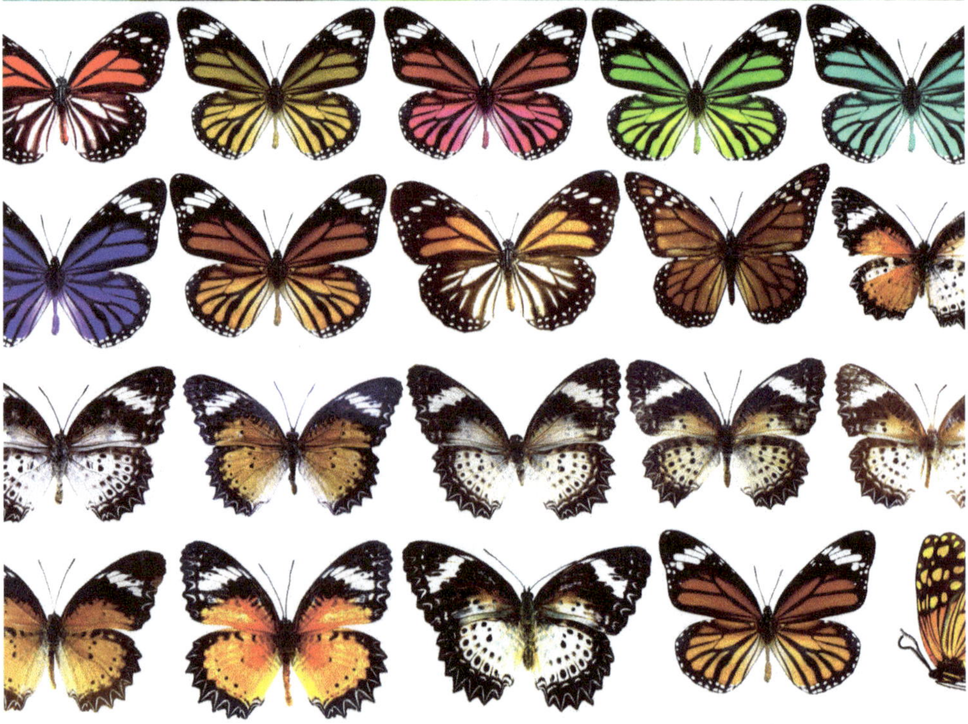

	**Vol.1** For All People who wish to take care of Climate Change, Food Industries: (Meat, Beverage, Dairy, Bakeries, Tortilla, Grain and Oilseed, Fruit and Vegetable, Seafood, And Sugar and Confectionery)
	**Vol.2** For All People who wish to take care of Climate Change, Electrical Industries: (Renewable Energy, Biofuels, Solar Heating & Cooling, Hydroelectric Power, Solar Power, Wind Power, Energy Conservation, Geothermal and Nuclear Power)
	**Vol.3** For All People who wish to take care of Climate Change, Fashion & Textile Industries: (Fashion Design, The Fashion System, Fashion Retailing, Marketing and Marchandizing, Textile Design and Production, Clothing and Textile Recycling)
	**Vol.4** For All People who wish to take care of Climate Change, Health & Beauty Industries: (Fragrances, Makeup, Cosmetics, Personal Care, Sunscreen, Toothpaste, Bathing, Nailcare & Shaving, Skin Care, Foot Care, Hair Care and Other Health & Beauty Products)

	# Vol.5 For All People who wish to take care of Climate Change, Maintenance & Cleaning Products: (All-purpose Cleaners, Abrasive Cleaners, Powders. Liquids, Specialty Cleaners, Kitchen, Bathroom, Glass and Metal Cleaners, Bleaches, Disinfectants and Disinfectant Cleaners)
	# Vol.6 For All People who wish to take care of Climate Change, Wood & Stationery Industries: (Wooden Products, Cardboard, Papers, Markers, Pens, NoteBooks, Writing Pads and Writing Sets, Pencils, White Papers, Envelopes and Organizers, Staplers and Paper Clips)
	# Vol.7 For All People who wish to take care of Climate Change, DIY & Construction Industries: (Do it yourself" ("DIY") of Building, Modifying, or Repairing, Renovation, Construction Materials, Cement, Coarse Aggregates. Clay Bricks, Power Cables, Pipes and Fittings, Plywood, Tiles, Natural Flooring)
	# Vol.8 For All People who wish to take care of Climate Change, Agricuture & Gardening Industries: (Shifting Cultivation, Nomadic Herding, Livestock Ranching, Commercial Plantations, Mixed Farming, Horticulture, Butterfly Gardens, Container Gardening, Demonstration Gardens, Organic Gardening)

	## Vol.9 For All People who wish to take care of Climate Change, Professional Products & Services: (Teachers, Pilots, Lawyers, Advertising Professionals, Architects, Accountants, Engineers, Consultants, Human Resources Specialist, R&D, Psychologists, Pharmacist, Commercial Banker, Research Analyst)
	## Vol.10 For All People who wish to take care of Climate Change, Financial Products & Services: (Banking, Professional Advisory, Wealth Management, Mutual Funds, Insurance, Stock Market, Treasury/Debt Instruments, Tax/Audit Consulting, Capital Restructuring, Portfolio Management)
	## Vol.11 For All People who wish to take care of Climate Change, Tourism Industries: (Airline Industry, Travel Agent, Car Rental, Water Transport, Coach Services, Railway, Spacecraft, Hotels, Shared Accommodation, Camping, Bed & Breakfast, Cruises, Tour Operators)
	## Set Box Books Vol.1-11 + Free Knowledge Test for Schools, Libraries, Homes and Offices all over the globe: www.TopTenAward.Net